ROCK ON!

*The Ultimate Guide to Cool Rocks
and Minerals for Kids*

Holly Creach

CONTENTS

Title Page

Copyright

Introduction   1

Are You A Rockhound?   3

The Three Types of Rocks   14

The Rock Cycle   35

Classifying Rocks   48

Fascinating Fossils: Windows to the Past   61

The Marvelous World of Metals   68

Gemstones: Earth's Magical Jewels   76

Agates, Crystals, and Geodes Oh My!   89

The Story of Fossil Fuels   96

Jobs and Careers for Rockhounds   102

Acknowledgements   111

INTRODUCTION

Rocks are the very foundation of our planet - the ancient building blocks that have shaped the landscapes, cultures, and history of the Earth. Whether you are a curious child, an outdoor adventurer, or a lifelong collector, the allure of discovering unique and beautiful rock, mineral, and fossil specimens is one that has captivated humans for millennia.

This comprehensive guide, will be your companion as you embark on an exploration of the geological world. From the basics of rock identification to the intriguing stories behind the formation of our planet, these chapters will equip you with the knowledge and skills to become an expert rockhound.

You'll start by learning about the diverse individuals who have caught the 'rock hound' bug - from young students to world-renowned geologists. Discover the thrilling history of rock collecting, tracing it back to the pioneering work of natural philosophers and indigenous cultures. Understand the special appeal that draws people to this hobby, whether it's the joy of discovery, the pursuit of scientific knowledge, or the simple connection to the natural world.

Next, dive into the fundamental classifications of rocks - igneous, sedimentary, and metamorphic - and explore the dynamic rock cycle that governs their transformation over time. Armed with this geological foundation,

Venture even deeper into the Earth's secrets as you uncover the incredible world of fossils - windows into our planet's distant past. Learn how these preserved remnants of ancient life can reveal insights about prehistoric environments, mass extinctions, and the evolution of species.

Fascination awaits as you discover the valuable metals and gemstones that have captivated human civilizations for centuries. Explore the intriguing stories behind these natural wonders,

Whether your passion lies in collecting stunning agates and crystal clusters or unearthing the mysteries of ancient life preserved in shale and sandstone, this book will be your indispensable guide. Prepare to be amazed by the incredible diversity and beauty of our planet's geological treasures

So grab your rock hammer, your magnifying glass, and your sense of wonder, and get ready to embark on an extraordinary journey.

ARE YOU A ROCKHOUND?

Introduction

Have you ever wondered about people who love to collect rocks? Do you like collecting pretty treasures off the ground? If you do, then you are a rockhound! Many adventurers like yourself search far and wide for all sorts of interesting rocks, minerals, and fossils. Let's dive into the exciting world of rockhounding and learn about these treasure hunters!

Who Hunts for Rocks?
Rockhounds come from all kinds of backgrounds. Kids, parents, scientists, and adventurers all love searching for rocks. Anyone with a curious mind and a love for nature can be a rockhound! Here are some examples of who hunts for rocks:

- **Kids**: Many children love collecting rocks because they come in so many cool shapes, colors, and sizes. It's like finding treasure!
- **Parents and Families**: Families often go rockhounding together as a fun outdoor activity. It's a great way to spend time together and explore nature.
- **Scientists**: Geologists and paleontologists study rocks and fossils to learn more about Earth's history. They often hunt for rocks to discover new information. ● **Hobbyists**: Lots of people collect rocks just for fun. They might polish them, turn them into jewelry, or simply enjoy looking at their collections.
- **Teachers and Students**: Sometimes, teachers and students go on field trips to collect rocks and learn more about geology.
- **Adventurers**: People who love exploring new places often hunt for rocks as part of their adventures. They might find interesting rocks while hiking, camping, or traveling.

No matter who you are, if you enjoy finding and learning about rocks, you can be a rockhound too!

Who Were the First Rockhounds?

It's hard to say exactly who the very first rockhound was because people have been interested in rocks, minerals, and fossils for thousands of years, but we can talk about some of the earliest people who loved studying and collecting rocks.

Early Rock Collectors

Georgius Agricola (1494-1555): Georgius Agricola was a scientist from Germany who loved studying rocks and minerals. He wrote a big book all about mining these materials called "De Re Metallica." This book was one of the first to explain how people find and use different rocks.

Nicholas Steno (1638-1686): Nicholas Steno was a scientist from Denmark who studied fossils and rocks. He figured out important ideas about how layers of rock form and change over time. His work helped people understand Earth's history better.

William Smith (1769-1839): William Smith was an English geologist who made the first detailed map of the rocks and minerals in England. His work was like creating a giant treasure map of all the different kinds of rocks in the country.

Ancient Cultures and Indigenous Peoples

Long before these scientists, many ancient cultures and indigenous peoples collected and used rocks and minerals. They made tools, jewelry, and other important items from these materials. For example:

- **Ancient Egyptians** used colorful minerals to make paint and makeup.
- **Native American tribes** used rocks like flint to make

tools and arrowheads.

- **Chinese** and **Mesoamerican cultures** carved beautiful artwork and important objects from stones like jade.

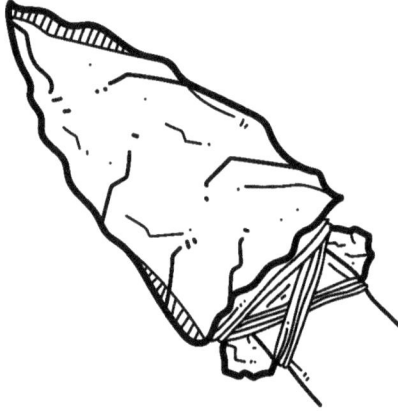

The First Hobby Rockhounds

If we think about rockhounding as a fun hobby, it became more popular in the 1800s and early 1900s. People started collecting rocks just for fun and to learn more about them. They even formed clubs and societies to share their discoveries and learn from each other.

While we can't say exactly who the very first rockhound was, we know that many people throughout history have loved collecting and studying rocks. From ancient cultures to early scientists, and later, hobbyists, the fascination with rocks has been around for a very long time. Whether they used rocks to make tools, studied them to understand the Earth, or collected them just for fun, people have always been excited by the beauty and mystery of rocks.

How Did Rock Hunters Get the Name Rockhound?
Imagine a hound dog sniffing around, following scents and searching for hidden treasures. That's kind of like what rockhounds do! The name "rockhound" came about because people who love to find pretty treasures off the ground are like hounds (or dogs) sniffing around for rocks instead of sniffing for scents.

Why Do They Hunt for Rocks?

There are many reasons why people love rockhounding:

Discovery

Rockhounds love the thrill of finding new rocks, minerals, and fossils. Every find is like discovering a hidden treasure! It's exciting to see what you'll uncover next.

Learning

Rockhounding is a great way to learn about Earth's history and the different types of rocks and minerals. It's like being a detective, solving mysteries from the past. You get to learn how rocks are formed and what they're made of.

Nature Connection

Being out in nature and exploring different landscapes is one of the best parts of rockhounding. It's a chance to enjoy the fresh air, listen to the sounds of nature, and see beautiful places. It's like going on an adventure in the great outdoors.

Creativity

Some rockhounds use the rocks they find to create beautiful artwork, like jewelry or sculptures. It's amazing how something as simple as a rock can become a piece of art! You can paint them, polish them, or turn them into cool crafts.

Community

Rockhounds often share their discoveries with others who love rocks too. They join clubs, go on trips together, and attend rock shows to meet new friends who share their passion. It's fun to talk about your finds and learn from others.

Where Do They Hunt for Rocks?

Rockhounds explore all sorts of places in search of rocks:

Beaches and Coastlines

Sandy shores and rocky cliffs are great spots to find colorful sea glass, smooth pebbles, and interesting shells washed up by the waves. You never know what the ocean might bring to the shore.

Mountains and Hills

Climbing mountains and hiking through hills can lead to discovering unique rocks and minerals hidden in the ground. These places are full of surprises waiting to be found.

Deserts

The vast desert landscapes hold treasures like crystals, geodes, and fossils waiting to be found beneath the sandy dunes. Deserts can be full of hidden gems.

Forests and Rivers

Exploring forests and riversides might reveal sparkling quartz crystals, colorful gemstones, or ancient fossils. The flowing water of rivers can uncover interesting rocks along the banks.

Quarries and Mines

Rockhounds are people who love collecting and studying

rocks, minerals, and fossils. Sometimes, they visit quarries and mines to find cool and rare rocks. Quarries are big open pits where rocks are dug up, while mines are tunnels that go deep underground.

These places are special because they have minerals and crystals that are hard to find anywhere else. For example, rockhounds might find shiny crystals like quartz, colorful stones like garnet, or other amazing minerals. Each trip to a quarry or mine is like a treasure hunt!

When rockhounds go to these places, they need to be very careful. They wear safety gear like helmets, gloves, and strong shoes to protect themselves. Sometimes, they need special permission to explore these sites. Following the rules helps keep everyone safe and protects the environment.

Even though it can be challenging, finding a beautiful, rare rock or crystal makes the adventure exciting and fun for rockhounds. It's a great way to learn about nature and discover hidden treasures beneath the ground!

Rockhounding is a exciting hobby that lets people explore the wonders of our planet. Whether you're searching for shiny crystals, fossilized creatures, or smooth pebbles, there's always something exciting to discover when you're a rockhound. The thrill of finding something new, learning about the Earth, enjoying nature, being creative, and making new friends all make rockhounding an amazing adventure.

So grab a magnifying glass, put on your hiking shoes, and join the fun world of rock collecting! Whether you're just starting out or have been collecting for a while, there's always something new and exciting to find.

THE THREE TYPES OF ROCKS

Rocks are everywhere. Valleys, rivers, mountains, oceans and even the atmosphere have rocks! There are three main types: igneous, sedimentary, and metamorphic. Each type of rock forms in different ways and has its own special features. Let's explore these types and learn how they are made.

Igneous Rocks - Baby Rocks Born from FIRE!
Igneous rocks are born from fiery, molten rock which comes from under the earth's crust that cools and hardens. This molten rock is called magma when it's underground and lava when it erupts from a volcano. This is how all rocks are born. Igneous rocks can be classified into two groups, intrusive and extrusive. Let's discover more about these two types of baby rocks.

Intrusive Igneous Rocks

Intrusive igneous rocks are special rocks that form deep underground from molten rock called magma. This happens when magma cools and hardens very slowly beneath the Earth's surface. Because it takes a long time for the magma to cool, these rocks end up with large crystals that you can see easily.

How Are They Formed?

Imagine a big pool of hot, melted rock deep underground. As this magma cools down very slowly over thousands or even millions of years, it starts to harden. Since it cools so slowly, the minerals in the magma have plenty of time to form large crystals. These rocks are called intrusive because they intrude, or push into, existing rocks below the Earth's surface.

What Do They Look Like?

Intrusive rocks have large, visible crystals that make them look quite different from rocks that cool quickly. Here are some common examples:

1. **Granite**: Granite is a very common intrusive

rock. It's usually light in color with big crystals of quartz, feldspar, and mica. You might see granite used for kitchen countertops and buildings because it's very strong and looks nice.

2. **Diorite**: Diorite has a mix of light and dark minerals, giving it a speckled appearance. It's not as common as granite but is still used for construction and decorative stones.
3. **Gabbro**: Gabbro is a dark, dense rock with large crystals. It forms from magma that is rich in iron and magnesium.

Why Are They Important?

Intrusive igneous rocks are important because they help us understand what happens deep inside the Earth. They are also useful in everyday life. For example, granite is used in buildings and monuments because it is durable and attractive.

So, the next time you see a granite countertop or a stone building, you'll know that it's made from rock that formed deep underground over millions of years!

Extrusive Rocks

Extrusive rocks are a special type of rock that forms from lava. Lava is molten rock that erupts from a volcano. When this hot, liquid rock comes out and cools quickly on the Earth's surface, it turns into extrusive rocks. These rocks are also called volcanic rocks because they are born from volcanoes!

How Are They Formed?

Extrusive rocks form when lava cools and hardens quickly after a volcanic eruption. Because the cooling happens so fast, there isn't much time for large crystals to grow. Instead, the rocks have tiny crystals that are often too small to see without a microscope.

What Do They Look Like?

Extrusive rocks have a few different looks. Here are some common types.

Basalt: This is the most common extrusive rock. It's dark in color, usually black or gray, and has a smooth texture. Basalt makes up most of the ocean floor.

Pumice: Pumice is a light, airy rock full of holes. It's so light it can float on water! Pumice is used in beauty products to scrub off dead skin.

Obsidian: Obsidian looks like shiny black glass. It forms when lava cools very quickly. Ancient people used obsidian to make sharp tools and weapons because it breaks into very sharp pieces, just like glass!

Why Are They Important?

Extrusive rocks are important for many reasons. They help scientists understand how volcanoes work and what happens during eruptions. Some extrusive rocks are used in construction, like basalt, or for making tools, like obsidian.

So, the next time you hear about a volcano, remember the incredible rocks that are born from its fiery eruption!

Sedimentary Rocks - Layers of History (Old rocks recycled into new rocks)

Sedimentary rocks are a cool type of rock that forms from bits and pieces of other rocks, minerals, plants, and even animals. These pieces are called sediments, and they get pressed and stuck together over time to form layers of sedimentary rocks. Kind of like a cake!

How Are They Formed?

Imagine bits of sand, pebbles, mud, and even tiny shells getting carried by wind or water to a new place, like the bottom of a river, lake, or ocean. These sediments settle in layers, one on top of another. Over many years, the layers get buried and squished together. Minerals in the water act like glue, sticking the sediments together to form solid rock.

What Do They Look Like?

Sedimentary rocks often have layers, which can look like stripes. Here are some common examples:

Sandstone: Sandstone is made from sand particles that have been cemented together. It can be red, yellow, or brown and often looks grainy. You might find sandstone in deserts or beaches

Shale: Shale is made from tiny mud particles and feels smooth. It's usually dark and can easily split into thin sheets. Shale forms in places like lake bottoms where the water is calm.

Limestone: Limestone is often made from the remains of tiny sea creatures. It can be white, gray, or even have fossils in it. Limestone is used to make cement and can also be found in many buildings and statues.

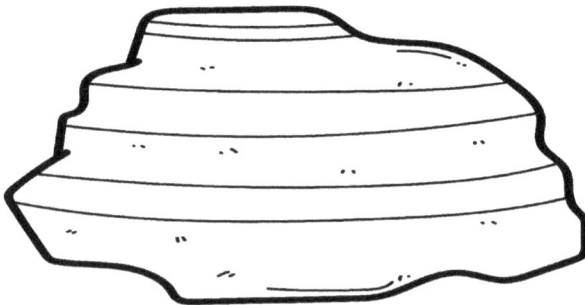

Why Are They Important?

Sedimentary rocks are important because they often contain fossils, which are the remains of ancient plants and animals. These fossils help scientists learn about life on Earth millions of years ago. Sedimentary rocks are also useful in construction and are sources of important resources like coal and oil.

Next time you see a layered rock, remember it might have started as tiny bits of sand or mud, slowly turning into rock over millions of years!

Metamorphic Rocks - The Transformers (igneous or sedimentary in disguise!)

Metamorphic rocks are a special type of rock that started as another kind of rock but changed because of intense heat and pressure deep inside the Earth. If igneous are baby rocks, and sedimentary are old recycled rocks, then these guys are flashy teenagers! Imagine baking a cake: you start with raw ingredients, but after heating them in the oven, they turn into something completely different. That's what happens with metamorphic rocks, and maybe teenagers as well!

How Are They Formed?

Metamorphic rocks form when existing rocks, called parent rocks, are buried deep underground. The heat from the Earth's interior and the pressure from the layers of rock above cause the minerals in the parent rocks to change. This process is called metamorphism and can take millions of years.

What Do They Look Like?

Metamorphic rocks often have unique textures and patterns. Here are some common examples:

- **Slate**: Slate comes from shale, a type of sedimentary rock. It's usually dark and splits into thin, flat sheets. People use slate for roofing, chalkboards, and tiles.
- **Marble**: Marble forms from limestone. It's smooth, often white, and can be polished to a shiny finish. Marble is used for statues, buildings, and countertops. It's so pretty that artists love to carve it!
- **Gneiss**: Gneiss (pronounced "nice") has a banded or striped appearance with alternating layers of light and dark minerals. It forms from rocks like granite. Gneiss is strong and used in construction.

Why Are They Important?

Metamorphic rocks are important because they help scientists understand the conditions deep within the Earth. They are also very useful in everyday life. For example, marble and slate are used in buildings and art.

Next time you see a marble statue or a slate tile, remember that these rocks have been through an incredible transformation deep underground!

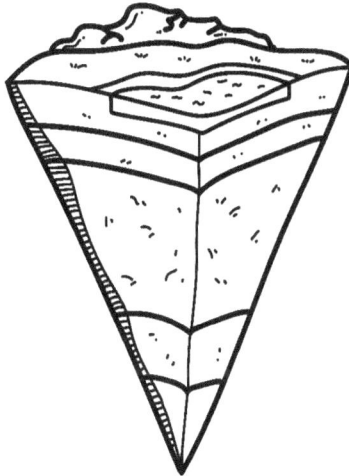

Fun Activity: Rock Detective Adventure

Are you ready to become rock detectives and learn about the three types of rocks? Let's embark on a fun adventure where we'll explore igneous, sedimentary, and metamorphic rocks. We'll do some cool experiments and activities to understand how each type of rock is formed. Before starting any experiment, make sure you have an adult to help you. Grab your magnifying glass, put on your explorer hat, and let's get started!

Activity: Creating Your Own Igneous Rocks

Objective: Understand how igneous rocks form from molten rock (magma/lava).

Materials:

- Chocolate chips (representing magma/lava)
- A microwave-safe bowl
- A spoon
- A microwave
- A plate
- A refrigerator

Steps:

1. **Melt the Chocolate**: Place a handful of chocolate chips in the microwave-safe bowl. Ask an adult to help you melt the chocolate in the microwave until it becomes smooth and liquid, like magma.
2. **Pour the "Magma"**: Pour the melted chocolate onto the plate. Spread it out a little bit to see what happens when magma cools quickly.
3. **Cool Quickly**: Place the plate with the melted chocolate in the refrigerator. This represents magma cooling quickly on the Earth's surface to form extrusive igneous rocks like basalt.
4. **Observe**: After the chocolate has hardened, take it out of the refrigerator. Look closely at the chocolate to see how it has formed a smooth, solid "rock." This is your own extrusive igneous rock!
5. **Cool Slowly**: For a second experiment, let some of the melted chocolate cool slowly at room temperature. This represents magma cooling slowly underground to

form intrusive igneous rocks like granite.

6. **Compare**: Compare the quickly cooled chocolate to the slowly cooled chocolate. Notice how the texture and appearance differ, just like real igneous rocks.

Activity: Making Sedimentary Rocks

Objective: Learn how sedimentary rocks form from layers of sediments.

Materials:

- A clear plastic cup or jar
- Sand
- Small pebbles
- Soil
- Water
- A spoon

Steps:

1. **Layer the Sediments**: Start by putting a layer of sand at the bottom of your plastic cup or jar. This represents one layer of sediment.
2. **Add Pebbles**: Next, add a layer of small pebbles on top of the sand. This is another type of sediment.

3. **Add Soil**: Add a layer of soil on top of the pebbles. These different layers represent how sediments build up over time.
4. **Add Water**: Pour a little water into the cup or jar to help compact the sediments. This simulates the natural process of compaction and cementation that turns sediments into sedimentary rocks.
5. **Press Down**: Use a spoon to press down on the layers to compact them further.
6. **Wait and Observe**: Let the layers settle and dry for a day or two. Once they are dry, look closely at how the layers have formed. This is similar to how sedimentary rocks like sandstone and shale are created.

Activity: Creating Metamorphic Rocks

Objective: Understand how heat and pressure transform rocks into metamorphic rocks.

Materials:

- Colored modeling clay (at least two colors)
- A rolling pin or sturdy cylinder
- A warm place (like a sunny windowsill or near a heater)

Steps:

1. **Prepare the Clay**: Take two different colors of modeling clay and flatten each piece into a thin layer.
2. **Layer the Clay**: Place one layer of clay on top of the other to create a "sedimentary" layer.
3. **Apply Pressure**: Use a rolling pin to press down on the layers of clay, applying pressure. This simulates the

pressure that forms metamorphic rocks.

4. **Add Heat**: Place the pressed layers of clay in a warm place for a few hours. This simulates the heat that helps transform rocks into metamorphic rocks.

5. **Observe the Changes**: After the clay has warmed up, look closely at how the layers have fused together and changed. This represents how sedimentary rocks can become metamorphic rocks like slate or marble.

Rock Detective Notebook

As a rock detective, it's important to record your findings. Create a rock detective notebook where you can:

1. **Draw Pictures**: Draw pictures of your chocolate "igneous rocks," layered sedimentary jar, and pressed clay "metamorphic rocks."

2. **Write Observations**: Write down what you noticed during each activity. How did the rocks change? What did they look like?

3. **Collect Samples**: If you can, find real examples of igneous, sedimentary, and metamorphic rocks outside. Glue small samples or pictures into your notebook and label them.

Sharing Your Discoveries

Share what you've learned with friends and family. Explain how you created your own rocks and what each type of rock is like. You can even put on a "Rock Detective Show" and demonstrate the activities to others.

By doing these fun activities, you've learned a lot about the three types of rocks and how they form. Igneous rocks come from melted rock, sedimentary rocks are made from layers of sediments, and metamorphic rocks are transformed by heat and pressure. Keep exploring and discovering the amazing world of rocks around you. Who knows, you might find a new favorite hobby or even become a real rock scientist someday! Happy exploring, rock detectives!

THE ROCK CYCLE
Nature's Amazing Recycling System

Now that you have learned the 3 types of rocks, igneous, sedimentary and metamorphic, have you ever wondered where rocks come from and why they look different? Just like water goes through a cycle of evaporation, condensation, and precipitation, rocks go through their own special cycle. This is called the rock cycle. Let's dive in and explore this awesome process!

What Are Rocks again?

Before we start, let's review what rocks are. Rocks are solid materials that make up the Earth's crust. They are made up of minerals, which are natural, non-living substances. You see rocks everywhere: in your backyard, at the beach, and even in mountains. There are three main types of rocks, each formed in different ways. These are igneous (baby rocks born from fire), sedimentary (old rocks recycled into layers), and metamorphic rocks (teenagers transformed from by heat, pressure and time!).

Weathering, Erosion, and Deposition: How Rocks Change Over Time

Have you ever noticed how mountains, rocks, and even sidewalks can slowly break down and change over time? This happens because of three natural processes: weathering, erosion, and deposition. Let's dive in and explore these amazing processes that shape our world!

Weathering Breaks it!

Weathering is the process where rocks are broken down into smaller pieces. There are two main types of weathering: physical weathering and chemical weathering.

Physical Weathering

Physical weathering happens when rocks are broken into smaller pieces without changing their chemical makeup or composition. Here are some ways physical weathering occurs:

1. **Frost Wedging**: When water seeps into cracks in rocks and then freezes, it expands and makes the cracks bigger. Over time, this can break the rock apart.
2. **Thermal Expansion**: Rocks expand when they are heated by the sun and contract when they cool down at night. This constant expanding and contracting can cause the rocks to crack and break apart.
3. **Biological Activity**: Plants and animals can also cause physical weathering. For example, tree roots can grow into cracks in rocks and force them apart as they grow. Animals digging and burrowing can also break rocks into smaller pieces.
4. **Abrasion**: Wind, water, and ice can carry particles that scrape against rocks, wearing them down like sandpaper. This can smooth out and round the edges of rocks over time.

Chemical Weathering

Chemical weathering occurs when the minerals in rocks change due to chemical reactions. Here are some common types of chemical weathering:

Dissolution: Some minerals in rocks can dissolve in water, especially if the water is slightly acidic. This process can cause the rock to break down and form new shapes.

Oxidation: When minerals in rocks react with the oxygen, they can form new compounds. For example, when iron in rocks reacts with oxygen, it forms rust, which weakens the rock.

Hydrolysis: This happens when minerals in rocks react with water. The water can change the minerals into different substances, causing the rock to weaken and break apart.

Erosion Takes It!

Once rocks are broken down by weathering, the next step is erosion. Erosion is the process of moving the small pieces of rock, called sediments, from one place to another. Here are the main agents of erosion:

Water Erosion

Water is a powerful force that can carry sediments far away from their original location.

Rivers and Streams: Flowing water in rivers and streams can pick up sediments and carry them downstream. The faster the water moves, the more sediment it can carry.

Rainfall: Rain can cause erosion by washing away loose sediments on the ground. Heavy rains can create small channels, called rills, that can grow into larger gullies over time.

Ocean Waves: Waves crashing against the shore can erode rocks and cliffs, carrying the sediments out to sea or along the coast.

Wind Erosion

Wind can also carry small particles of rock and soil over long distances.

Deflation: Wind can blow away loose sediments from the surface, leaving behind larger particles and creating areas of bare rock or desert pavement.

Abrasion: Windblown sediments can act like sandpaper, wearing away rocks and structures they hit. This can create smooth surfaces and interesting rock formations.

Ice Erosion

Ice, especially in the form of glaciers, can move huge amounts of rock and sediment.

Glacial Movement: Glaciers are massive, slow-moving rivers of ice. As they move, they pick up rocks and sediments, carrying them along. When the glacier melts, it drops these materials in new locations.

Plucking: Glaciers can freeze onto rocks and pull them out of the ground as they move. This can create jagged landscapes and deep valleys.

Deposition Drops it!

After sediments are transported by erosion, they eventually settle down in a new location. This process is called deposition. Deposition can create various landforms and is an important part of the rock cycle. Here's how deposition works with different agents:

Water Deposition

River Deltas: When a river slows down as it enters a larger body of water, like an ocean or a lake, it drops its sediments. These sediments can build up and form a delta, which is a landform that looks like a triangle or a fan.

Alluvial Fans: In dry regions, when a stream comes down a mountain and slows down at the base, it can drop its load of sediments and create an alluvial fan.

Beaches: Waves can deposit sand and other sediments along the shore, creating beaches.

Wind Deposition

Sand Dunes: In deserts or along beaches, wind can deposit sand to form dunes. These hills of sand can move and change shape over time as the wind continues to blow.

Loess: Fine particles of dust and silt can be carried by the wind and deposited over large areas. These deposits, called loess, can create fertile soil that's great for farming.

Ice Deposition

Moraines: When glaciers melt, they leave behind piles of rocks and sediments called moraines. These can form ridges or hills along the sides and ends of where the glacier once moved.

Drumlins: These are smooth, elongated hills formed by glacial deposits. They show the direction the glacier was moving.

The Big Picture: How It All Fits Together

Weathering, erosion, and deposition are like a team that works together to shape our landscape. Here's how these processes are connected:

Weathering: Rocks break down into smaller pieces through physical and chemical weathering.

Erosion: These pieces are picked up and moved by water, wind, or ice.

Deposition: The sediments are eventually dropped in a new location, creating new landforms.

These processes happen very slowly, often over thousands or even millions of years. But if you look closely, you can see their effects all around you. The sand on the beach, the soil in your garden, and the shape of mountains and valleys are all results of weathering, erosion, and deposition. Here's a little rhyme to help you remember:

Weathering takes
it, Erosion breaks
it, and when the
motion stops it,
Deposition drops it!

Fun Activities to Learn More

Here are some fun activities you can try to see these processes in action (make sure you have an adult to help you):

Weathering Experiment: Take a piece of chalk and place it in a cup of vinegar. Watch as the vinegar (acid) breaks down the chalk, similar to how chemical weathering works on rocks.

Erosion in a Pan: Fill a shallow pan with sand or soil. Create a slope by propping one end of the pan up. Pour water slowly from the top and watch how it carries the sand or soil down the slope, mimicking water erosion.

Make a Sand Dune: On a windy day, go to a sandbox or a sandy area. Use a small fan to blow the sand and watch how it forms dunes, similar to wind deposition.

Weathering, erosion, and deposition are incredible processes that shape the Earth's surface. They work together to break down rocks, move sediments, and create new landforms. By understanding these processes, we can better appreciate the ever-changing landscape around us. So next time you're outside, take a moment to observe the rocks, soil, and landforms and think about the amazing journey they've been on!

CLASSIFYING ROCKS

How Scientists Classify Rocks

Have you ever picked up a rock and wondered what kind it is or how it was formed? Rocks are all around us, and each one has its own story to tell. Scientists use various characteristics to classify rocks and understand their origins. Today, we'll explore how scientists classify rocks based on texture, color, hardness, density, luster, minerals, and the acid test. Let's dive in!

Texture: The Look and Feel of Rocks

The texture of a rock refers to the size, shape, and arrangement of its grains or crystals. Texture tells us a lot about how the rock was formed.

Examples of Rock Texture

- **Coarse-Grained**: Rocks with large, easily visible crystals. These rocks, like granite, form when magma cools slowly underground.
- **Fine-Grained**: Rocks with small, hard-to-see crystals. Basalt is an example and forms from lava that cools quickly on the Earth's surface.
- **Glassy**: Rocks with no visible crystals, very smooth. Obsidian forms when lava cools extremely quickly.
- **Clastic**: Rocks made of fragments of other rocks, typical in sedimentary rocks like sandstone.
- **Foliated**: Rocks with a layered or banded appearance due to pressure, found in metamorphic rocks like gneiss.

Activity: Texture Hunt

Objective: Identify different rock textures.

Materials: Notebook, pencil, magnifying glass.

Steps:

1. Go on a walk and collect various rocks (make sure to ask a responsible adult before you go).
2. Use the magnifying glass to observe each rock's texture.
3. Write down your observations. Are the grains large or small? Is the rock smooth or layered?

Color: The Hues of Rocks

The color of a rock can tell us about the minerals it contains. Different minerals give rocks different colors.

Common Rock Colors:

Light-Colored: Often contain minerals like quartz and feldspar. Example: Granite.

Dark-Colored: Can contain minerals like pyroxene and olivine. Example: Basalt.

Varied Colors: Sedimentary and metamorphic rocks can have a mix of colors depending on their mineral content.

Fun Activity:

Color Observation

Objective: Explore the colors

of rocks.

Materials: Notebook, colored pencils.

Steps:

1. Collect a variety of rock (remember to ask a responsible adult first).
2. Draw each rock in your notebook and color them to match their real colors.
3. Try to guess what minerals might be in each rock based on its color.

Hardness: Scratch and See

Hardness measures how easily a rock can be scratched. Scientists use the Mohs scale, which ranges from 1 (softest) to 10 (hardest). The Mohs scale is a way to see how hard different minerals are by scratching them with other things. Here's a list of some common minerals on the Mohs scale, with examples and simple ways to test their hardness:

Talc (1)
- Example: Talcum powder (like baby powder).
- Testing: You can easily scratch talc with your fingernail. It feels really soft and slippery.

Gypsum (2)
- Example: Chalk.
- Testing: You can scratch gypsum with your fingernail, but it's a bit harder than talc. It can also be scratched by a penny.

Calcite (3)
- Example: Some seashells and limestone.
- Testing: You can scratch calcite with a penny, but not with your fingernail. It's harder than gypsum.

Quartz (7)
- Example: Some beach sand and many colorful crystals.
- Testing: Quartz is very hard. It can scratch glass and steel, like the blade of a knife.

Diamond (10)
- Example: Jewelry diamonds.
- Testing: Diamond is the hardest mineral. It can scratch anything, even other diamonds.

How to Test Mineral Hardness

1. **Fingernail (2.5 on the Mohs scale):**

   O   Try to scratch the mineral with your fingernail. If it leaves a mark, the mineral is softer than your fingernail.

2. **Penny (3.5 on the Mohs scale):**

   O   Use a penny to scratch the mineral. If it leaves a mark, the mineral is softer than the penny.

3. **Steel Knife (5.5 on the Mohs scale):**

   O   Carefully use the edge of a knife to scratch the mineral. If it leaves a mark, the mineral is softer than the knife.

4. **Glass Plate (5.5 on the Mohs scale):**

   O   Try to scratch a piece of glass with the mineral. If it leaves a scratch, the mineral is harder than the glass.

Safety Tips

- Always have an adult help you when using sharp objects or glass.
- Be careful not to scratch surfaces that aren't meant to be scratched.
- Wear safety glasses to protect your eyes.

Fun Facts

- Minerals can have different hardness even if they look the same because they might have little bits of other things in them.
- The Mohs scale isn't like a ruler; the steps between numbers are not equal. For example, diamond is much harder than quartz, even though they are just three numbers apart on the scale.

Fun Activity: Hardness Test

Objective: Test the hardness of different rocks.

Materials: Rock samples, copper penny, steel nail, piece of glass.

Steps:

1. Try to scratch each rock with the copper penny. Does it leave a mark?

2. If not, try the steel nail, then the piece of glass.
3. Record your findings and rank the rocks from softest to hardest.

Density: How Heavy Is It?

Density is how heavy a rock feels for its size. Rocks with closely packed minerals are denser and feel heavier.

Fun Activity: Density Comparison

Objective: Compare the density of different rocks.

Materials: Rock samples, balance scale, water, measuring cup.

Steps:

1. Weigh each rock and record its mass.
2. Fill a measuring cup with a known amount of water, maybe use a container with a scale on the side, and submerge each rock.
3. Measure how much the water level rises (this tells you the volume).
4. Compare the mass to the volume to understand which rocks are denser.

Luster: Shiny or Dull?

Luster describes how a rock reflects light. It can be shiny, glassy, or dull.

Types of Luster:

1. **Metallic**: Shiny like metal. Example: Pyrite (fool's gold).
2. **Vitreous**: Glassy shine. Example: Quartz.
3. **Dull**: Doesn't reflect much light. Example: Chalk.

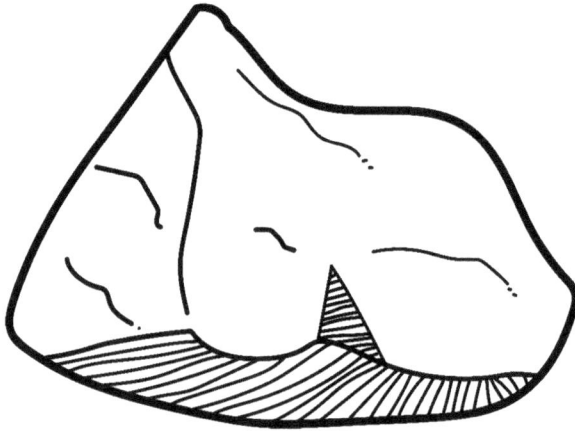

Fun Activity: Luster Observation

Objective: Observe the luster of different rocks.

Materials: Rock samples, flashlight.

Steps:

1. Shine the flashlight on each rock and observe how the light reflects.
2. Record whether the rock is metallic, glassy, or dull.

Minerals: The Building Blocks of Rocks

Minerals are the ingredients that make up rocks. Each mineral has its own unique properties.

Common Rock-Forming Minerals:

Quartz: Hard, glassy mineral found in many rocks.

Feldspar: Common in igneous rocks, usually light-colored.

Mica: Has a shiny, layered appearance, found in igneous and metamorphic rocks.

Calcite: Common in sedimentary rocks like limestone, reacts with acid.

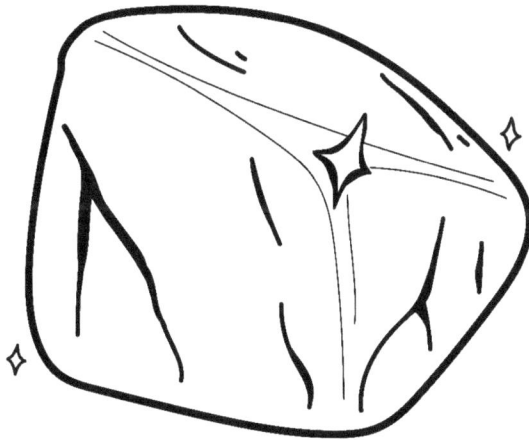

Fun Activity: Mineral

Identification Objective:

Identify minerals in rocks.

Materials: Rock samples, mineral guidebook, magnifying glass.

Steps:

1. Observe each rock closely with the magnifying glass.
2. Use the guidebook to identify the minerals based on color, shape, and other properties.
3. Write down which minerals you find in each rock.

Acid Test: Fizz and Fun

Some rocks react with acids. This test helps identify rocks containing calcium carbonate, like limestone.

Fun Activity: Acid Test

Objective: See how rocks react to acid.

Materials: Rock samples, vinegar, dropper, safety goggles.

Steps:

1. Put on your safety goggles.
2. Place a few drops of vinegar on each rock.
3. Observe what happens. Does the rock fizz or bubble? If it does, it might be limestone or another carbonate rock.
4. Record your observations.

Putting It All Together: Classifying Rocks

Now that we know about texture, color, hardness, density, luster, minerals, and the acid test, let's put all this knowledge together to classify rocks.

Steps to Classify a Rock:

1. **Observe the Texture**: Is it coarse-grained, fine-grained, glassy, clastic, or foliated?
2. **Check the Color**: Is it light, dark, or varied?

3. **Test the Hardness**: Where does it fall on the Mohs scale?
4. **Compare the Density**: How heavy does it feel for its size?
5. **Look at the Luster**: Is it shiny, glassy, or dull?
6. **Identify the Minerals**: What minerals can you see?
7. **Do the Acid Test**: Does it react with vinegar?

Example: Classifying a Mystery Rock

Let's classify a mystery rock together!

1. **Texture**: The rock has large crystals. (Coarse-grained)
2. **Color**: It's light-colored with spots of white and pink. (Light)
3. **Hardness**: It scratches glass but not a steel nail. (Around 6-7 on Mohs scale)
4. **Density**: It feels heavy for its size. (Dense)
5. **Luster**: It has a glassy shine. (Vitreous)
6. **Minerals**: You can see quartz, feldspar, and mica.
7. **Acid Test**: No reaction with vinegar.

Based on these characteristics, our mystery rock is most likely **granite**, an igneous rock.

Classifying rocks is like being a detective. By looking at texture, color, hardness, density, luster, minerals, and using the acid test, we can uncover the secrets of each rock. Rocks have amazing stories to tell about how they were formed and what they've been through. So next time you pick up a rock, try to classify it using these characteristics. Who knows what fascinating story you'll uncover! Happy rock hunting!

FASCINATING FOSSILS: WINDOWS TO THE PAST

Introduction to Fossils

Have you ever wondered what life was like millions of years ago, long before humans walked the Earth? Fossils are like time capsules that give us a glimpse into that ancient world. They are the remains or traces of plants, animals, and other organisms that lived long ago. Let's explore the exciting world of fossils and learn how they form, where to find them, and why they are important.

What Are Fossils?

Fossils are the preserved remains or traces of living things from the past. These remains can be bones, teeth, shells, leaves, or even footprints. Fossils help scientists, called paleontologists, understand what life was like on Earth long ago.

How Do Fossils Form?

Fossils form in several ways, but the most common process is called fossilization. Here are the steps of how a typical fossil forms:

1. **Living Organism:** An animal or plant lives and then dies.
2. **Rapid Burial:** The remains get buried quickly by mud, sand, or other materials, protecting them from scavengers and decay.

3. **Sediment Accumulation:** Over time, more layers of sediment cover the remains, and pressure builds up.
4. **Mineralization:** Minerals in the water seep into the buried remains, replacing the organic material with rock-like minerals.
5. **Exposure:** Over millions of years, the sediment becomes rock, and the fossil may be exposed by erosion or excavation.

Types of Fossils

There are several types of fossils, each providing different information about ancient life:

1. **Body Fossils:** These are the actual remains of an organism, such as bones, teeth, and shells. For example, dinosaur bones and mammoth tusks are body fossils.
2. **Trace Fossils:** These are marks or evidence left by an organism, like footprints, burrows, or feces. Trace fossils can tell us about the behavior of ancient animals.
3. **Molds and Casts:** A mold fossil forms when an organism is buried in sediment and then decays, leaving an empty space. A cast fossil forms when that space gets filled with minerals, creating a replica of the original organism.
4. **Amber Fossils:** Sometimes, small creatures like insects get trapped in tree resin. Over time, the resin hardens into amber, preserving the trapped organisms in amazing detail.
5. **Petrified Wood:** This happens when plant material is buried by sediment and then replaced by minerals, turning it into stone while retaining its original structure.

Famous Fossils

Some fossils are so famous that they have become household names. Here are a few examples:

1. **Lucy:** Lucy is one of the most famous fossils ever discovered. She is an Australopithecus afarensis, an early human ancestor who lived about 3.2 million years ago. Her discovery

helped scientists understand human evolution.

2. **Tyrannosaurus Rex (T. Rex):** The T. Rex is one of the most well-known dinosaurs. Its fossils have been found in North America, and it was one of the largest meat-eating dinosaurs.
3. **Trilobites:** Trilobites are extinct marine creatures that lived hundreds of millions of years ago. Their fossils are found all over the world and are often very detailed.
4. **Archaeopteryx:** This fossil is a link between dinosaurs and birds. It had feathers like a bird but also teeth and a long tail like a dinosaur.

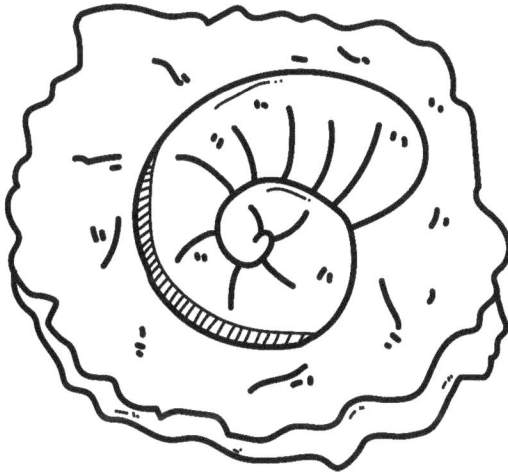

Where to Find Fossils

Fossils can be found in many places, from deserts to riverbeds. Here are some common locations where fossils are discovered:

Sedimentary Rock Layers: Most fossils are found in sedimentary rocks, which form from layers of sediment. These rocks can be found in riverbeds, cliffs, and quarries.

Deserts: Many famous dinosaur fossils have been found in desert regions, where erosion exposes ancient rock layers.

Beaches: Fossils can sometimes be found along coastlines where waves and erosion reveal ancient layers of rock.

Caves: Some fossils are found in caves, where they have been protected from the elements for millions of years.

Road Cuts and Construction Sites: Sometimes, fossils are discovered during construction work when digging into the ground exposes hidden rock layers.

How to Become a Fossil Hunter

Do you want to become a fossil hunter? Here are some tips to get you started:

Research Locations: Find out where fossils have been found near you. Look for public sites where fossil hunting is allowed.

Use the Right Tools: Basic tools include a small hammer, chisel, brush, and safety goggles. Always be careful when using these tools.

Look for Clues: Fossils are often found in sedimentary rocks. Look for layers of rock that might contain fossils, and be patient. Fossil hunting takes time and careful observation.

Ask for Permission: Always get permission before hunting for fossils on private land or in protected areas.

Respect Nature: Leave the site as you found it, and don't damage any fossils you find. If you discover something significant, report it to a museum or local paleontologist.

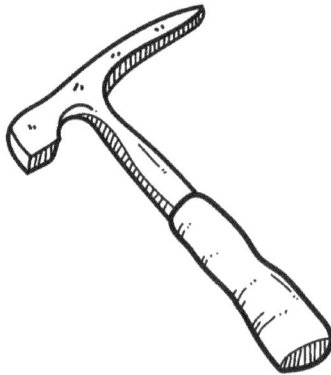

Why Fossils Are Important

Fossils are not just cool to look at; they are incredibly important for understanding our planet's history. Here are a few reasons why fossils matter:

Learning About the Past: Fossils help scientists learn about the plants and animals that lived long ago and how they interacted with their environment.

Evolution: By studying fossils, scientists can trace how different species have changed over time. This helps us understand the process of evolution and how life on Earth has developed.

Climate Analysis: Fossils can provide clues about ancient climates. For example, finding tropical plant fossils in Antarctica tells us that the climate there was once much warmer.

Geological History: Fossils help geologists understand the history of the Earth's surface, including the movement of continents and the formation of mountains and oceans.

Fun Facts About Fossils

- **Oldest Fossils:** The oldest known fossils are about 3.5 billion years old and are microscopic bacteria.
- **Fossilized Poop:** Called coprolites, fossilized feces (poop) can provide information about an ancient animal's diet.
- **Dinosaur Eggs:** Fossilized dinosaur eggs have been found, some with embryos still inside. These fossils give us clues about dinosaur reproduction and development.
- **Giant Fossils:** The largest dinosaur fossil ever found belongs to Argentinosaurus, which is estimated to have been over 100 feet long.

How to Start Your Own Fossil Collection

Collecting fossils can be a fun and educational hobby. Here are some tips to get you started:

1. **Start Small:** Begin with easily accessible fossils like seashells or plant impressions. You can find these in many places, including local parks or beaches.

2. **Label Your Finds:** Keep track of where and when you found each fossil. This makes your collection more interesting and valuable.
3. **Learn to Identify Fossils:** Use a guidebook or online resources to help identify the fossils you find. Look at their shape, size, and other features.
4. **Store Them Safely:** Keep your fossils in a safe place where they won't get damaged. Displaying them in a case or on a shelf is a great way to show them off.
5. **Join a Club:** Many areas have fossil or rock clubs where you can meet other fossil enthusiasts, go on field trips, and learn more about fossils.

Fun Activity: Make Your Own Fossil Imprint

Here's a fun activity to help you understand fossils better!

What You Need:

- Modeling clay
- Small toy dinosaur or plastic leaf
- Rolling pin
- Paint (optional)

Instructions:

1. Roll out the modeling clay until it's flat and smooth.
2. Press the toy dinosaur or plastic leaf into the clay to make an imprint.
3. Carefully remove the toy or leaf.
4. Let the clay dry completely.
5. Optional: Once dry, you can paint your fossil imprint to make it look more realistic.

Now you have your very own fossil imprint! This helps you imagine how real fossils form over millions of years.

Fossils are amazing windows into the past that help us understand the history of life on Earth. From the tiniest ancient bacteria to the mighty T. Rex, each fossil has a story to tell. By learning about fossils, we can appreciate the incredible diversity of life that has existed on our planet and the processes that have shaped it over millions of years. So next time you pick up a rock, take a closer look – you might just find a piece of history hidden inside!

THE MARVELOUS WORLD OF METALS

What Are Metals?

Imagine you have a treasure chest filled with shiny, hard objects that you can shape into anything you want. These magical treasures are called metals. Metals are materials found in the Earth's crust, and they are known for being shiny, solid, and strong. Some common metals you might know include iron, copper, gold, and silver. These metals are used in many ways to make our lives easier and more exciting!

How Does the Earth Make Metals?

The Earth has been making metals for billions of years. It's a bit like baking a cake, but way more intense and much hotter! Deep inside the Earth, there are special processes that create metals. Here's how it works:

1. **Earth's Ingredients**: The Earth started with lots of different elements. These elements are the building blocks of everything around us, like iron, copper, and gold.
2. **Heat and Pressure**: Deep within the Earth, it's incredibly hot and there's a lot of pressure. This heat and pressure cause chemical reactions that form metals. Imagine a giant pressure cooker, but instead of cooking food, it's cooking up metals!
3. **Cooling Down**: Over time, the molten (melted) metal moves closer to the Earth's surface and starts to cool down. As it cools, it solidifies into metal deposits.

4. **Rocks and Ores**: These metal deposits get mixed with other rocks and minerals, creating ores. Ores are rocks that contain enough metal to be worth mining.

How Are Popular Metals Like Gold and Silver Mined?

Imagine you're on a treasure hunt, but instead of finding a pirate's chest, you're digging for shiny metals like gold and silver! Mining these precious metals is an exciting adventure that involves several cool steps.

First, geologists (rock scientists) search for places where gold and silver are likely hiding. They use maps, special tools, and even satellite images to find the best spots. It's like using a treasure map! Once they find a promising location, it's time to dig!

There are two main types of mining:

1. **Open-Pit Mining:** This method is used when gold or silver is close to the surface. Miners remove large amounts of soil and rock to create a big, open pit. It looks like a giant hole in the ground!
2. **Underground Mining:** When gold and silver are buried deep underground, miners dig tunnels and shafts to reach them. Imagine exploring a maze of tunnels, but instead of finding cheese, you find shiny metals!

After the ores (rocks containing gold or silver) are dug up, they need to be processed to get the precious metals out. This involves crushing the ore into tiny pieces and using chemicals or heat to separate the gold or silver from the rock. It's like squeezing the last bit of toothpaste from the tube!

Finally, the purified gold and silver are melted and shaped into bars, coins, or jewelry. These shiny treasures are then ready to be used or admired!

So, mining gold and silver is a bit like a high-tech treasure hunt, with lots of digging, crushing, and melting to uncover those glittering gems!

How have metals shaped our world?

Humans have been using metals for thousands of years, and these shiny treasures have changed the course of history!

The Copper Age

About 5,000 years ago, humans discovered copper, one of the first metals they could easily shape. Copper is soft and easy to work with, making it perfect for early tools and ornaments. People in ancient times used copper to make knives, jewelry, and even mirrors!

The Bronze Age

Soon after the Copper Age, humans learned that mixing copper with tin made a stronger metal called bronze. This discovery led to the Bronze Age. Bronze was used to make better tools, weapons, and even the first metal coins. It was like upgrading from a simple wooden stick to a super-sophisticated gadget!

The Iron Age

Then came the Iron Age. Iron is much stronger than copper or bronze and can be sharpened to make very effective tools and weapons. Iron plows made farming easier, and iron swords changed the way battles were fought. This period saw great advances in technology and civilization.

How Humans Use Metals Today

Today, metals are everywhere! They play a huge role in our everyday lives, making things stronger, faster, and more reliable. Here are some ways we use metals today:

- **Building and Construction**: Skyscrapers, bridges, and houses all rely on metals like steel (which is made from iron) and aluminum to be strong and durable.

Transportation: Cars, buses, trains, and airplanes are made from various metals. These metals make vehicles strong, light, and fast.

Technology: Your favorite gadgets, like smartphones, computers, and tablets, all contain metals. These metals are crucial for conducting electricity and making the devices work.

Household Items: Everyday objects like forks, spoons, and even some toys are made from metals. These items are sturdy and long-lasting thanks to metals.

Medical Uses: Metals are used in medical equipment and even in the human body. For example, titanium is used for hip replacements because it's strong and doesn't rust.

Metals as Currency: Shiny Money!

Metals have been used as currency for thousands of years. Let's dive into the history of how metals became money and what metals are found in our coins today.

Early Metal Money

Before paper money and digital payments, people used metals as money. Metals were durable, easy to carry, and had intrinsic value (they were valuable just by being themselves). Here's how it all started:

Bronze and Copper Coins: The earliest metal coins were made of bronze and copper. These metals were abundant and easy to shape into coins.

Gold and Silver Coins: As civilizations grew wealthier, they started using gold and silver for coins. These precious metals were highly valued and perfect for making valuable coins. The shiny, glittery nature of gold and silver made them popular choices for currency.

Metals in Modern Currency

Today, we still use metals in our coins, but they're often a mix of different metals to make them more durable and cost-effective. Here are some common metals found in coins:

Copper: Many coins are made primarily of copper or have a copper core. For example, U.S. pennies are made of zinc with a thin copper coating.

Nickel: This metal is used in many coins because it's strong and resistant to corrosion. Nickels, dimes, and quarters in the U.S. contain a mix of copper and nickel.

Silver: While pure silver coins are rare today, many coins used to be made of silver. Some special commemorative coins are still made of silver.

Gold: Gold coins are now mostly collectible or used for investment. They are beautiful and valuable, but not practical for everyday transactions.

Fun Facts About Metal Money

- **The Ancient Greeks and Romans**: They used coins made of bronze, silver, and gold. The value of the coin was based on the metal's weight and purity.
- **The First Silver Dollar**: The U.S. introduced the silver dollar in 1794. It was made of 90% silver!
- **Gold Rush Money**: During the California Gold Rush, people minted coins directly from gold nuggets they found.

The Importance of Recycling Metals

Recycling metals is super important for our planet. Metals can be recycled over and over again without losing their properties. When we recycle metals, we save energy and reduce the need to mine new ores, which helps protect the environment.

For example, recycling aluminum saves 95% of the energy needed to make new aluminum from ore. It's like giving old soda cans a new life, turning them into something useful again and again.

The Future of Metals

As we look to the future, metals will continue to play a crucial role in our lives. Scientists are developing new technologies to make mining and processing metals more sustainable and environmentally friendly. They are also finding innovative ways to use metals, like in medicine for making implants and in renewable energy systems.

Imagine a world where metals help power our homes with clean energy from the sun and wind, or where new materials made from metals revolutionize the way we live and work. The possibilities are endless!

Metals are incredible materials that have been part of human history for thousands of years. From ancient copper tools to modern skyscrapers and smartphones, metals have shaped our world in countless ways. Understanding how the Earth makes metals, where they are found, and how we use them today helps us appreciate their value and work towards using them responsibly.

Next time you hold a coin, a fork, or a shiny piece of jewelry, remember the amazing journey it took from deep within the Earth to your hands. And who knows, maybe one day you'll be a scientist, engineer, or treasure hunter, discovering new ways to use and recycle these marvelous materials!

Fun Activity: Mining for Iron using a Magnet!

Get ready for a fun and educational activity that uses a magnet to find hidden iron in the ground! This activity will help you learn about how magnets work and how iron is found in nature.

Materials Needed:

- A strong magnet (a neodymium magnet works best, but any strong magnet will do)
- A piece of string or chord.
- A plastic bag or small container

- A notebook and pencil for taking notes
- A small garden trowel or shovel (optional)

Preparation:

1. **Safety First**: Make sure to do this activity in a safe area, such as your backyard, a park, or a garden. Always have an adult with you to supervise.
2. **Understanding Magnets**: Before you start, learn a bit about how magnets work. Magnets attract certain metals, like iron, because of their magnetic field. This activity will help you see this in action!

The Hunt:

1. **Prepare Your Magnet**: Tie the string or chord to your magnet. This will allow you to drag the magnet on the ground.
2. **Choose Your Spot**: Find a good spot in your yard or garden to start your search. Areas with loose soil or sand are ideal.
3. **Start Searching**: Slowly walk around your selected area dragging the magnet along the ground as you go. If there are any iron particles or small pieces of iron in the soil, they will be attracted to the magnet and stick to it.
4. **Check Your Finds**: After moving the magnet around for a while, check to see if any iron particles are sticking to it. Be sure to check the magnet often, removing any iron material you find and putting it into your plastic bag.
5. **Repeat**: Try searching in different areas to see where you find the most iron. You might be surprised at how much you can collect!

Analyze Your Finds:

1. **Observe**: Look closely at the iron particles you've collected. Use a magnifying glass if you have one to see them better.
2. **Record Your Results**: In your notebook, write down where

you found the most iron. Did you find more in one area than another? What do the iron particles look like?

Fun Facts About Iron:

- **Iron is the most common metal found in the Earth's crust.**
- **Iron is used to make steel, which is used in buildings, cars, and many other things.**
- **Meteorites often contain iron, so if you find a big piece of iron, it might have come from space!**

Bonus Activity: Iron Experiment

1. **Clean the Iron Particles**: Fill a bowl with water and gently swish the iron particles around to clean them. Let them dry on a paper towel. Once they are clean and dry, place them on a plate and put Ketchup on them. Wait for an hour, then wash it off. What happened? Research: Find out why Ketchup can clean iron.
2. **Magnetic Art**: Use your magnet and iron particles to create magnetic art! Place the magnet under a piece of paper and sprinkle the iron particles on top. Move the magnet around and watch the iron particles dance and create patterns.

This magnetic treasure hunt is a fun way to explore and learn about iron in the ground. You'll get hands-on experience with magnets and discover how iron is found in nature. Plus, you'll have a great time searching for hidden treasures in the soil. So grab your magnet and start your adventure – who knows what amazing iron finds you'll uncover!

GEMSTONES: EARTH'S MAGICAL JEWELS

Introduction to Gemstones

Gemstones are like the jewels in nature's crown. They come in all colors, shapes, and sizes, and they've been treasured by humans for thousands of years. There are two main types of gemstones: precious and semi-precious. Let's dive into the sparkling world of these amazing stones and learn all about them!

Precious Gemstones

Precious gemstones are the rare and highly valuable gems that everyone loves. There are four main types of precious gemstones:

Diamonds

- **Appearance**: Diamonds are usually clear and sparkling, but they can also come in colors like blue, pink, and yellow. They are famous for their brilliant sparkle.
- **Formation**: Diamonds are formed deep within the Earth under extreme pressure and heat. They are made of carbon atoms arranged in a crystal structure.
- **Where Found**: Diamonds are found in countries like South Africa, Russia, and Canada.
- **Mining**: Diamonds are mined from deep underground or from riverbeds. Miners dig through layers of rock to reach the diamond-rich areas.
- **Uses**: Diamonds are often used in jewelry, especially in engagement rings. They are also used in industrial tools for cutting and grinding because they are the hardest natural material on Earth. Diamonds can cut anything, even other diamonds!

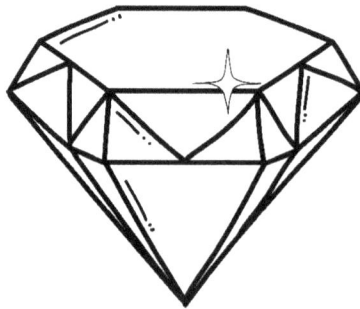

Emeralds

- **Appearance:** Emeralds are known for their rich green color. The greener and clearer the emerald, the more valuable it is.
- **Formation**: Emeralds form in hydrothermal veins, where hot water with minerals flows through cracks in rocks. This process takes millions of years.
- **Location**: The best emeralds come from Colombia, but they are also found in Brazil and Zambia.
- **Mining**: Emeralds are usually mined in tunnels dug into mountains. Miners carefully extract the emeralds from the surrounding rock.
- **Uses**: Emeralds are used in fine jewelry and are often associated with luxury and royalty. Famous historical figures, like Cleopatra, adored emeralds.

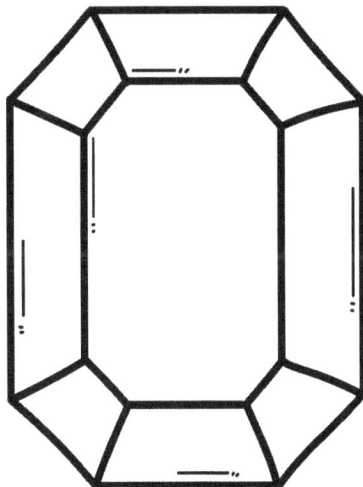

Sapphires

- **Appearance**: Sapphires come in many colors, but the most famous are deep blue.
  Sapphires can also be pink, yellow, green, and even white.
- **Formation**: Sapphires form in metamorphic rocks, remember the transformers? These are rocks that have been changed by heat and pressure.

- **Location:** Sapphires are found in places like Sri Lanka, Myanmar, and Madagascar.
- **Mining:** Sapphires are often mined from gravel beds in rivers. Miners sift through the gravel to find the precious stones.
- **Uses:** Sapphires are used in fine jewelry and are symbols of wisdom and nobility. Blue sapphires are especially prized for their rich color.

Rubies

- **Appearance:** Rubies are typically a vibrant red, which comes from the presence of chromium. The redder the ruby, the more valuable it is.
- **Formation:** Rubies form in metamorphic rocks, like sapphires. The process takes millions of years under high pressure and temperature.
- **Location:** The finest rubies come from Myanmar, but they are also found in Thailand and Sri Lanka.
- **Mining:** Rubies are mined in a similar way to sapphires, often from gravel beds in rivers or from underground mines.
- **Uses:** Rubies are used in fine jewelry and are symbols of passion, love, and courage. They are often called the "king of gemstones."

Semi-Precious Gemstones

Semi-precious gemstones are also beautiful and colorful, but they are more common and generally less expensive than precious gemstones. Here are some popular semi-precious gemstones:

Turquoise

- **Appearance**: Turquoise is known for its lovely blue-green color.
- **Formation**: Turquoise forms in arid regions where groundwater with copper and aluminum flows through rocks.
- **Location:** Turquoise is found in places like the southwestern United States, Iran, and Egypt.
- **Mining**: Turquoise is usually mined from shallow pits or tunnels.
- **Uses**: Turquoise is often used in jewelry and decorative items. Native American cultures have used turquoise for centuries in their jewelry and art.

Malachite

- **Appearance**: Malachite has striking green bands and patterns.
- **Formation**: Malachite forms near the surface of the Earth in the oxidized zones of copper deposits.
- **Location:** Malachite is found in countries like the Democratic Republic of Congo, Zambia, and Russia.
- **Mining**: Malachite is mined from shallow pits and tunnels.
- **Uses**: Malachite is used for jewelry, sculptures, and decorative objects. Its unique patterns make each piece one of a kind.

Opals

- **Appearance:** Opals are famous for their play of colors, showing flashes of green, blue, red, and yellow.
- **Formation**: Opals form from silica-rich water that seeps into

cracks in rocks and then evaporates, leaving behind silica deposits.

- **Location:** The best opals come from Australia, but they are also found in Ethiopia and Mexico.
- **Mining**: Opals are mined from open pits and underground tunnels.
- **Uses**: Opals are used in jewelry, especially rings and pendants. Their colorful play of light makes them very attractive.

Tiger Eye

- **Appearance**: Tiger eye has a unique golden-brown shimmer that looks like a tiger's eye.
- **Formation**: Tiger eye forms from quartz that has replaced fibrous minerals like crocidolite. This process creates the chatoyant, or cat's eye, effect.
- **Location:** Tiger eye is found in South Africa, Australia, and the United States.
- **Mining**: Tiger eye is usually mined from shallow pits.
- **Uses**: Tiger eye is often used in beads, bracelets, and rings. Its distinctive shimmer makes it popular for jewelry.

The Most Famous Gemstone in the World: The Hope Diamond

The Hope Diamond is one of the most famous precious gemstones in the world. This dazzling blue diamond weighs 45.52 carats and has a rich history filled with mystery and intrigue.

Where It Was Found

The Hope Diamond was originally discovered in the Golconda mines in India over 350 years ago. It started as a larger diamond that was cut into smaller pieces, including the Hope Diamond.

Its History

The diamond has passed through many hands over the centuries.

It was purchased by King
Louis XIV of France in 1668 and became part of the French crown
jewels, known as the "French Blue." During the French Revolution,
it was stolen and later reappeared in London, where it was recut
into its current size.

In the 19th century, it was bought by Henry Philip Hope, giving it
the name we know today. Over the years, it was owned by several
wealthy individuals and was rumored to be cursed, bringing bad
luck to its owners.

Where It Is Today

Today, the Hope Diamond is housed in the Smithsonian National Museum of Natural History in Washington, D.C. It's one of the most visited exhibits in the museum, admired by millions of people each year for its incredible beauty and fascinating history.

How Much Is The Hope Diamond Worth?

The value of the Hope Diamond is hard to pin down exactly, but experts estimate it to be worth around $250 million. This value comes not only from its size and stunning blue color but also from its rich history and the many legends surrounding it.

The Largest Gemstone in the World: The Cullinan Diamond

The biggest gemstone in the world is the Cullinan Diamond, discovered in South Africa in 1905. This diamond was a whopping 3,106 carats before it was cut, which is about the size of a small melon!

Its History

The diamond was named after Sir Thomas Cullinan, who owned the mine where it was found. It was such a big deal that it was sent to England in a plain box to keep it safe from thieves. Once in England, it was presented to King Edward VII as a gift.

The Cullinan Diamond was then cut into nine large gemstones and about 100 smaller ones. The largest piece, called the "Great Star of Africa," is 530 carats and is set in the Sovereign's Sceptre with Cross, a fancy staff that the British monarch holds during special events. Another piece, the "Lesser Star of Africa," is 317 carats and is part of the Imperial State Crown.

How Much Is It Worth?

Today, these pieces are kept in the Tower of London, where people can visit and see them. The exact value of the Cullinan Diamond pieces is hard to determine, but it's estimated to be worth over $2 billion! That's like having a treasure chest full of gold and jewels, making it the most expensive gemstone in the world!

Fun Activity: Create Your Own Gemstone Art

Now that you know all about precious and semi-precious gemstones, let's have some fun and create your own gemstone art!

What You'll Need

- Colored paper or cardstock
- Scissors
- Glue
- Markers or crayons
- Glitter (optional)
- A printed or drawn outline of a gemstone shape (like a diamond or emerald)

Steps

1. **Choose Your Gemstone**: Pick a gemstone you want to create, like a diamond, emerald, or sapphire.
2. **Cut Out the Shape**: Use the printed or drawn outline of the gemstone and cut it out from the colored paper or cardstock.
3. **Decorate Your Gemstone**: Use markers or crayons to color your gemstone. You can use glitter to add some sparkle if you like. Try to match the colors to the gemstone you chose. For example, use green for an emerald or blue for a sapphire.
4. **Create a Background**: Cut out another piece of colored paper to use as a background for your gemstone. You can decorate the background with patterns or designs that match your gemstone.
5. **Glue Your Gemstone**: Glue your decorated gemstone onto the background paper.
6. **Show Off Your Art**: Display your gemstone art in your room or give it to a friend or family member as a gift.

Discussion

- **Share Your Creation**: Show your gemstone art to your family and friends. Explain why you chose that particular gemstone and what makes it special.

- **Learn More**: Look up more information about your chosen gemstone. Where is it found? How is it formed? What are some famous pieces of jewelry that feature this gemstone?

Gemstones are incredible natural treasures that add beauty and sparkle to our lives. Whether they are precious gems like diamonds and emeralds or semi-precious stones like turquoise and opals, each gemstone has its own unique story and charm. By learning about how they are formed, where they are found, and how they are used, we can appreciate these wonderful gifts from nature even more. So next time you see a piece of jewelry or a sparkling gem, you'll know just how special it really is!

AGATES, CRYSTALS, AND GEODES OH MY!

Introduction

Rocks are full of wonders, especially when you discover the secrets inside agates, crystals, and geodes. These special rocks might look ordinary on the outside, but inside, they hide incredible beauty and fascinating stories. Let's discover how the earth makes these natural treasures, where they are found and which ones are famous!

Agates: Nature's Striped Marvels

What Are Agates?

Agates are a type of rock known for their beautiful, colorful bands. These bands can come in many different colors and patterns, making each agate unique. They are a type of chalcedony, which is a mineral made of quartz. People love agates for their stunning looks and sometimes even for their supposed magical properties!

How Are Agates Formed?

Agates start their journey in volcanic areas. When a volcano erupts, it spews out lava. As the lava cools, it forms bubbles or cavities. Over millions of years, water that contains minerals seeps into these bubbles. The minerals slowly deposit inside the cavity, layer by layer, forming the beautiful bands that agates are famous for. This metamorphic process takes a long time, but the result is worth the wait!

Where Are Agates Found?

Agates can be found all over the world, but some places are especially famous for them. You can find agates in countries like Brazil, Mexico, and the United States, particularly in states like Oregon and Montana. These places have lots of volcanic rocks where agates can form.

Famous Agates

Some agates are so beautiful and rare that they become very famous and expensive. For example, the "Lace Agate" from Mexico is known for its intricate, lacy patterns. Another famous type is the "Moss Agate," which looks like it has tiny plants trapped inside. These special agates can be worth a lot of money to collectors.

Crystals: Nature's Jewels

What Are Crystals?

Crystals are minerals that have formed into specific shapes. They grow in a repeating pattern, which makes them look like they have sharp edges and flat surfaces. Crystals can come in many different colors and sizes, and they are often very beautiful to look at.

How Are Crystals Formed?

Crystals form when liquid rock (magma) or mineral-rich water cools down and starts to harden. As the liquid cools, the minerals inside it come together and start to form crystals. This can happen deep underground or in cracks in rocks. The conditions need to be just right for crystals to grow, with the perfect temperature, pressure, and amount of minerals.

Where Are Crystals Found?

Crystals can be found in many places around the world. Some of the best places to find crystals are in countries like Brazil, Madagascar, and the United States. Crystals can also be found in caves, mines, and even in your own backyard if you're lucky!

Largest Crystals in the World! Crystal Cave

Deep beneath the surface of Mexico lies a magical place known as the Crystal Cave. Discovered in 2000 by miners working in the Naica Mine, this incredible cave is filled with some of the largest natural crystals ever found.

The crystals in this cave are made of selenite, a form of gypsum, and they are truly gigantic! Some of the crystals are over **30**

feet long and weigh as much as **55 tons**. Imagine a crystal taller than a house and heavier than several elephants combined! These crystals have been growing for hundreds of thousands of years, in an environment that is perfect for their formation.

Geodes: The Hidden Treasures

What Are Geodes?

Geodes are like nature's surprise packages. They look like regular, round rocks on the outside, but when you crack them open, they reveal a hollow space lined with beautiful crystals inside. Geodes can be small enough to fit in your hand or large enough to stand in like the Crystal Cave in Mexico.

How Are Geodes Formed?

Geodes start out as bubbles in volcanic or sedimentary rock. Over time, mineral-rich water seeps into these bubbles. As the water evaporates, the minerals are left behind, slowly forming crystals inside the hollow space. This process can take millions of years. The outside of a geode is usually made of limestone or another tough rock, while the inside is filled with crystals.

Where Are Geodes Found?

Geodes can be found all over the world, but they are most commonly found in deserts. You can find them in places like Mexico, Brazil, and the southwestern United States, especially in states like Arizona and California. Geodes are often found in areas with volcanic rock or limestone.

Famous and Geodes

Some geodes are very famous because of their size or the beauty of the crystals inside. The "Pulpí Geode" in Spain is one of the largest geodes ever found. It's big enough for people to walk inside and is filled with enormous, sparkling crystals. Another famous geode is the "Empress of Uruguay," which is filled with stunning purple amethyst crystals and is over 11 feet tall. These geodes are not only beautiful but also very valuable.

The Beauty and Uses of These Natural Wonders

Agates, crystals, and geodes are not just beautiful; they also have many uses. Here are some ways people use these natural wonders:

Jewelry and Decoration

One of the most common uses for agates, crystals, and geodes is in jewelry. Rings, necklaces, bracelets, and earrings made from these stones can be very beautiful and unique. People also use them to decorate their homes, creating stunning pieces that catch the light and add a touch of natural beauty to any room.

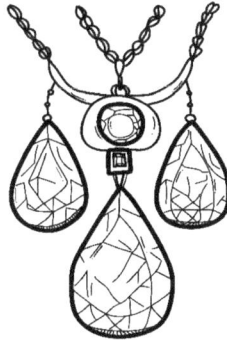

Healing and Meditation

Many people believe that crystals and agates have healing properties. They use them in meditation to help focus their mind and bring positive energy. For example, someone might hold a piece of amethyst while meditating to feel calm and centered, or place a rose quartz under their pillow to promote restful sleep and dreams of love and peace.

Collecting and Studying

Some people collect agates, crystals, and geodes as a hobby. Each stone is a piece of Earth's history, and collectors love to study them and learn about how they were formed. Museums often have large collections of these stones, showcasing their incredible variety and beauty.

Fun Facts to Impress Your Friends

- **Agates were used by ancient civilizations:** People have been using agates for thousands of years. Ancient Egyptians used them for jewelry and amulets, believing they had protective powers.
- **The largest geode ever found:** The Pulpí Geode in Spain is the largest geode ever discovered. It's big enough for people to walk inside and is filled with enormous crystals.
- **Quartz is all around us:** Quartz is one of the most common minerals on Earth. You can find it in sand, rocks, and even in the screens of your phones and computers!

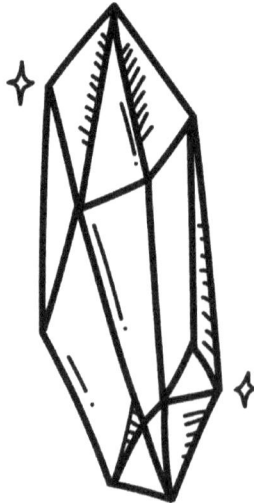

How to Start Your Own Collection

If you're inspired to start your own collection of agates, crystals, and geodes, here are a few tips to get you started:

1. **Learn about the different types:** Read books or watch videos about agates, crystals, and geodes to learn how to identify them.

2. **Visit a rock shop:** Local rock shops often have a wide variety of stones and knowledgeable staff who can help you pick out your first pieces.

3. **Go on a rock hunt:** Look for rocks in your backyard, local parks, or on vacation. Always make sure it's okay to collect rocks in the area you're exploring.

4. **Display your collection:** Find a special place to keep your collection, like a shelf or a display case, where you can enjoy looking at your stones every day.

The Magic Beneath Our Feet

Next time you pick up a rock, remember that there could be an incredible story hidden inside.

Agates, crystals, and geodes show us that the Earth is full of wonders waiting to be discovered. Whether you collect them, use them for healing, or simply admire their beauty, these stones are a reminder of the magic that lies beneath our feet. So keep exploring, and who knows what amazing treasures you might find!

THE STORY OF FOSSIL FUELS

Introduction to Fossil Fuels

Have you ever wondered what makes your car move or how your house stays warm during the winter? It all has to do with something called fossil fuels. These amazing natural resources have been powering our world for a long time though. Let's take a journey to learn about fossil fuels, how humans started using them, what we use them for today, how they are extracted and refined, their impact on the environment, and what the future holds for them.

The History of Fossil Fuels

What Are Fossil Fuels?

Fossil fuels are natural resources like coal, oil, and natural gas. They are called "fossil" fuels because they are made from the remains of ancient plants and animals that lived millions of years ago. Over millions of years, heat and pressure turned these remains into the fuels we use today.

How Did Humans Start Using Fossil Fuels?

A long time ago, people discovered that certain black rocks could burn and produce heat. These rocks were coal, and people began using them to heat their homes and cook their food. This was the start of humans using fossil fuels.

In the 18th century, during the Industrial Revolution, people found out that coal could be used to power steam engines. This was a big deal because steam engines could run factories, trains, and ships. Later, in the 19th century, people discovered oil and natural gas. These new fossil fuels could power even more things, like cars and airplanes.

What We Use Fossil Fuels for Today

Today, fossil fuels are everywhere. They power our cars, planes, and trains. They heat our homes and produce electricity. They are even used to make plastics, chemicals, and many other products we use every day.

Electricity Production

One of the biggest uses of fossil fuels is to produce electricity. Power plants burn coal, oil, or natural gas to create steam. This steam turns turbines, which generate electricity. This electricity powers our homes, schools, and gadgets.

Transportation

Fossil fuels like gasoline and diesel fuel cars, trucks, buses, and airplanes. Without fossil fuels, it would be much harder to travel long distances.

Heating

Many homes use natural gas or oil to stay warm in the winter. These fuels are burned in furnaces to produce heat.

Products

Fossil fuels are used to make many everyday products, such as plastics, medicines, and even clothing.

How Much Fossil Fuels Do We Use?

We use a lot of fossil fuels! Every day, people around the world consume millions of barrels of oil, tons of coal, and billions of cubic feet of natural gas. This huge amount of energy keeps our modern world running but also raises concerns about how long these resources will last and their impact on our planet.

Extracting Fossil Fuels from the Earth

Coal Mining

Coal is extracted from the earth through mining. There are two main types of mining: surface mining and underground mining. Surface mining is used when coal is close to the earth's surface. Huge machines scrape away the soil and rocks to get to the coal. Underground mining is used when coal is deep underground. Miners dig tunnels to reach the coal.

Oil Drilling

Oil is found deep underground or under the ocean floor. To extract it, companies use drilling rigs to drill holes and pump the oil to the surface. Sometimes, oil is found in rock formations, and special techniques like hydraulic fracturing, or "fracking," are used to release it.

Natural Gas Extraction

Natural gas is often found alongside oil. It is extracted using similar drilling methods. Once brought to the surface, it is processed to remove impurities before being sent through pipelines to homes and businesses.

Refining Fossil Fuels for Use

Once fossil fuels are extracted, they often need to be refined before they can be used.

Coal Processing

Coal is usually cleaned to remove dirt and other impurities before it is burned. This helps it burn more efficiently and reduces pollution.

Oil Refining

Oil refining is a complex process. Crude oil, which comes from the ground, is made up of many different types of hydrocarbons. Refineries heat the crude oil to high temperatures. As it heats, different parts of the oil evaporate and are collected separately. This process creates products like gasoline, diesel, jet fuel, and heating oil.

Natural Gas Processing

Natural gas is cleaned to remove impurities like water, carbon dioxide, and sulfur. It is then cooled and condensed into a liquid for easy transport or sent through pipelines as gas.

The Impact of Fossil Fuels on the Environment

While fossil fuels have many benefits, they also have a significant impact on the environment.

Air Pollution

Burning fossil fuels releases pollutants into the air. These pollutants can cause health problems and contribute to smog and acid rain.

Greenhouse Gas Emissions

Fossil fuels release carbon dioxide (CO_2) and other greenhouse gases when burned. These gases trap heat in the atmosphere and contribute to global warming and climate change.

Water Pollution

Oil spills and runoff from coal mines can pollute rivers, lakes, and oceans, harming wildlife and ecosystems.

Habitat Destruction

Mining and drilling can destroy habitats and disrupt ecosystems. Surface mining, in particular, can transform entire landscapes.

The Future of Fossil Fuels

As the world becomes more aware of the environmental impact of fossil fuels, people are looking for alternative energy sources. Renewable energy sources like solar, wind, and hydroelectric power are becoming more popular. These sources are cleaner and won't run out.

Transition to Renewable Energy

Many countries are investing in renewable energy to reduce their dependence on fossil fuels. Solar panels and wind turbines are becoming more common sights, and new technologies are being developed to make renewable energy even more efficient.

Energy Efficiency

Improving energy efficiency is another important step. This means using less energy to do the same job. For example, energy-efficient appliances and light bulbs use less electricity, and better insulation can reduce the need for heating and cooling.

Innovations in Fossil Fuel Use

There are also efforts to make fossil fuel use cleaner. Technologies like carbon capture and storage (CCS) can capture CO_2 emissions from power plants and store them underground, reducing their impact on the environment.

Fun Facts About Fossil Fuels

- **Coal Was Once a Plant:** Millions of years ago, coal was made up of plants that lived in swamps. Over time, they were buried and turned into coal.
- **Oil is Sometimes Called "Black Gold":** This nickname comes from its value and importance in the world.
- **Natural Gas is Odorless:** The smell you associate with natural

gas is actually added for safety, so you can detect leaks.

Fossil fuels have played a huge role in shaping our world. They power our lives in many ways, from the cars we drive to the electricity that lights our homes. However, they also have significant impacts on our environment. As we move forward, finding cleaner and more sustainable energy sources is important. By learning about fossil fuels, we can better understand our world and work towards a brighter, greener future. Keep exploring, and you'll discover even more amazing things about the world around you!

JOBS AND CAREERS FOR ROCKHOUNDS

Introduction

Now that you know so much about Rocks, how can you make money being a rockhound? Rocks, minerals, and fossils hold the secrets of the Earth, and there are many exciting careers for people who love to study them. These careers are perfect for rockhounds like you, who are fascinated by the natural world. Let's dive into the various careers you can pursue if you love rocks and minerals, and learn about what each job entails.

Geologist: The Earth Detective

What Does a Geologist Do?

Geologists are scientists who study the Earth, its materials, and the processes that shape it. They investigate rocks, minerals, fossils, and the Earth's structure to understand its history and predict future changes. Geologists often work outdoors, collecting samples and conducting fieldwork. They also work in laboratories, analyzing samples and using technology to study the Earth.

Types of Geologists

There are many types of geologists, each specializing in different aspects of the Earth:

- **Petrologists:** Study rocks and their origins.
- **Mineralogists:** Focus on minerals and their properties.

- **Paleontologists:** Investigate fossils and ancient life forms.
- **Seismologists:** Study earthquakes and the movements of the Earth's crust. • **Volcanologists:** Explore volcanoes and volcanic activity.

Where Do Geologists Work?

Geologists work in various places, including:

- **Universities and Research Institutions:** Conducting scientific research and teaching students.
- **Oil and Gas Companies:** Searching for natural resources like oil and natural gas.
- **Mining Companies:** Finding and extracting valuable minerals and metals.
- **Government Agencies:** Studying natural hazards and managing natural resources.
- **Environmental Organizations:** Assessing the impact of human activities on the Earth.

Paleontologist: The Fossil Hunter

What Does a Paleontologist Do?

Paleontologists study fossils to learn about the history of life on Earth. They examine the remains of plants, animals, and other organisms that lived millions of years ago. By studying fossils, paleontologists can reconstruct ancient ecosystems and understand how life has evolved over time.

Types of Fossils

Paleontologists study different types of fossils, such as:

- **Body Fossils:** Remains of an organism's body, like bones, teeth, and shells.
- **Trace Fossils:** Evidence of an organism's activities, like footprints, burrows, and nests.
- **Microfossils:** Tiny fossils that require a microscope to see, such as pollen and plankton.

Where Do Paleontologists Work?

Paleontologists work in various settings, including:

- **Museums:** Curating fossil collections and creating exhibits for the public.
- **Universities and Research Institutions:** Conducting research and teaching students.
- **Field Sites:** Excavating fossils and exploring ancient environments.
- **Government Agencies:** Studying fossils to inform conservation and land management decisions.

Mineralogist: The Crystal Expert

What Does a Mineralogist Do?

Mineralogists study minerals, their properties, and how they form. They identify and classify minerals based on their chemical composition and crystal structure. Mineralogists also explore the uses of minerals in industry, medicine, and technology.

Types of Mineralogists

There are several specializations within mineralogy, such as:

- **Gemologists:** Focus on gemstones and their properties.
- **Economic Mineralogists:** Study minerals that are valuable for industrial use, like metals and ores.
- **Environmental Mineralogists:** Investigate the impact of minerals on the environment and human health.

Where Do Mineralogists Work?

Mineralogists work in various places, including:

- **Universities and Research Institutions:** Conducting research and teaching students.
- **Mining Companies:** Finding and analyzing mineral deposits.

- **Gemstone Companies:** Evaluating and certifying gemstones.
- **Museums:** Curating mineral collections and creating exhibits for the public.
- **Environmental Organizations:** Studying the effects of minerals on ecosystems and human health.

Environmental Scientist: The Earth Protector

What Does an Environmental Scientist Do?

Environmental scientists study the environment and the impact of human activities on it. They work to protect natural resources, prevent pollution, and promote sustainable practices. Environmental scientists often work on projects related to water quality, air pollution, soil health, and biodiversity.

Types of Environmental Scientists

There are several specializations within environmental science, such as:

- **Ecologists:** Study ecosystems and the interactions between organisms and their environment.
- **Hydrologists:** Investigate water resources and their management.
- **Soil Scientists:** Study soil properties and their impact on agriculture and the environment.
- **Conservation Scientists:** Work to protect natural resources and promote sustainable land use.

Where Do Environmental Scientists Work?

Environmental scientists work in various settings, including:

- **Government Agencies:** Developing and enforcing environmental regulations.
- **Research Institutions:** Conducting scientific research and

monitoring environmental changes.
- **Environmental Organizations:** Advocating for conservation and sustainability.
- **Consulting Firms:** Advising businesses and communities on environmental practices.

Gemologist: The Gem Detective

What Does a Gemologist Do?

Gemologists study gemstones, their properties, and their origins. They identify, classify, and evaluate gemstones based on their color, clarity, cut, and carat weight. Gemologists also assess the value of gemstones and work to detect synthetic or treated stones.

Types of Gemstones

Gemologists work with a wide variety of gemstones, including:

- **Diamonds:** Known for their brilliance and hardness.
- **Rubies:** Valued for their deep red color.
- **Emeralds:** Recognized for their rich green hue.
- **Sapphires:** Available in many colors, with blue being the most famous.

Where Do Gemologists Work?

Gemologists work in various places, including:

- **Jewelry Stores:** Evaluating and selling gemstones and jewelry.
- **Gemological Laboratories:** Testing and certifying gemstones.
- **Auction Houses:** Appraising and selling high-value gemstones.
- **Museums:** Curating gemstone collections and creating

exhibits for the public. **Mining Engineer: The Resource**

Developer

What Does a Mining Engineer Do?

Mining engineers design and oversee the process of extracting minerals and metals from the Earth. They work to ensure that mining operations are safe, efficient, and environmentally responsible. Mining engineers also develop new technologies to improve mining practices.

Types of Mining Engineers

There are several specializations within mining engineering, such as:

- **Exploration Engineers:** Locate and assess mineral deposits.
- **Operations Engineers:** Manage day-to-day mining activities.
- **Reclamation Engineers:** Restore mined land to its natural state.

Where Do Mining Engineers Work?

Mining engineers work in various settings, including:
- **Mining Companies:** Planning and managing mining operations.
- **Consulting Firms:** Advising on mining projects and practices.
- **Government Agencies:** Regulating and inspecting mining activities.
- **Research Institutions:** Developing new mining technologies and methods.

Fun Facts About Rockhound Careers

- **Geologists on Mars:** Some geologists work with space agencies like NASA to study rocks on other planets. They help plan missions to Mars and analyze samples brought back from space.
- **Dinosaur Discoveries:** Paleontologists have discovered thousands of dinosaur species. Some new species are still being found today!
- **Gemstone Origins:** Many famous gemstones, like the Hope Diamond, have fascinating histories and legends associated with them.

Fun Activity: Start Your Own Rock Collection

You can begin your journey as a rockhound right at home! Here's a fun activity to get you started.

What You Need:

- A notebook for recording your finds
- Magnifying glass
- Small boxes or bags for storing rocks
- A field guide to rocks and minerals (you can find these at a library or bookstore) • Labels and markers

Instructions:

1. **Go on a Rock Hunt:** Explore your backyard, local parks, or nature trails to find interesting rocks. Look for different colors, shapes, and textures.
2. **Collect and Store:** Collect the rocks you find and store them in small boxes or bags. Make sure to label each one with the date and place you found it.
3. **Identify Your Rocks:** Use your field guide and magnifying glass to identify the types of rocks and minerals you've collected. Record your findings in your notebook.
4. **Create a Display:** Arrange your rocks on a shelf or in a display

case. You can group them by type, color, or location.
5. **Share Your Collection:** Show your collection to friends and family. Tell them about your finds and what you've learned about each rock.

There are many exciting careers for rockhounds who love exploring the natural world. Whether you dream of becoming a geologist, paleontologist, mineralogist, environmental scientist, gemologist, or mining engineer, there's a career that's perfect for you. Each of these careers offers the chance to uncover the Earth's secrets, protect the environment, and work with fascinating materials. So, keep exploring, collecting, and learning about rocks and minerals— you never know where your passion might take you!

ACKNOWLEDGEMENTS

The creation of this comprehensive guide to the world of rocks, minerals, and fossils would not have been possible without the invaluable contributions of many individuals and organizations.

First and foremost, we extend our heartfelt gratitude to the generations of geologists, mineralogists, paleontologists, and rockhounding enthusiasts whose tireless work and discoveries have advanced our understanding of the Earth's geological history. Their passion for uncovering the secrets hidden within rocks has paved the way for this book to come to life.

We would like to thank the curators, researchers, and educators at renowned institutions such as the Smithsonian National Museum of Natural History, the American Museum of Natural History, and the Geological Society of America. Their expertise, collections, and educational resources have been instrumental in shaping the content within these pages.

We are also deeply indebted to the indigenous cultures and traditional knowledge keepers who have long revered the Earth's natural treasures and shared their wisdom with the world. Their stewardship and respect for the land have helped preserve the wonder of rocks, minerals, and fossils for future generations.

Finally, a special thank you to the talented team of writers, editors, illustrators, and designers who have poured their heart and soul into crafting this book. Your dedication to bringing the "Science of Rocks" to life has been truly remarkable.

It is our sincere hope that this guide will ignite a deeper appreciation for the geological world, inspire countless new rockhounds, and contribute to the ongoing exploration and celebrating our planet's remarkable natural history.

www.ingramcontent.com/pod-product-compliance
Lightning Source LLC
Chambersburg PA
CBHW050511210326
41521CB00011B/2418